Contents

Preface

The Number Play Workbook Series is designed to expose young children to the beauty and benefits of the world of Mathematics. "Number Play" comes in three volumes: <u>Number Play Yeah</u>, recommended for ages 3 - 5, <u>Number Play Hurrah</u> - ages 6-7 and <u>Number Play This Way</u>, ages 7 - 8. Older children who have not mastered basic number concepts will find "Number Play" appealing.

"Number Play Yeah" is the first volume in the Number Play Series, and promises exciting exercises to stimulate and challenge the young mind. It is the beginning of an adventure in integrated child centred exercises incorporating music, writing, listening, speaking, art and science.

The aim is to motivate young children to have fun, while discovering number concepts necessary to sharpen their computation skills. Number Play creates building blocks which are required for a strong foundation in Mathematics.

Parents and teachers will welcome "Number Play Yeah!" which has practical <u>Teacher/ Parent Instruction Notes</u> as well as suggestions for creating a stimulating *Math Lab* to bring additional excitement to your child as he/she explores the world of numbers.

This workbook series was written to challenge, motivate and excite students as they engage in number exercises. Children learn through play, therefore, learning should include fun and amusement.

The exercises in *"Number Play"* afford the child the freedom to explore and make his own discoveries. As he counts, colours, interacts and make decisions, he becomes aware of his potential to take charge of his own learning.

The book is designed for students to decide on their own colour scheme, choose pictures and create their own pattern in most instances, as they discover the versatility of numbers.

Number Play Workbooks are ideal for Home School as they provide a level of flexibility which parents will welcome, not too rigid, just enough balance to allow for creativity while you play the role of facilitator.

The Number Play Series can also be used in the classroom as a supplemental tool for children to enhance their Mathematical skills. As pupils engage in the number play activities, they are on track to becoming independent thinkers and problem solvers. The exercises allow children to hone skills which are necessary in charting their course to becoming scientists and leaders in their respective fields.

Tools to enhance Number Play.

The following tools are necessary for children to take charge of their learning as they engage in number play. Before starting the Number Play Series be sure to select the following items to create a Math Lab.

Additional tools are listed in Parent/ Teacher Instruction Notes.

Cell phone	clock	number cards
Watch	calendar	bottle tops
Musical items	blocks	telephone
Calculator	sandbox	measuring cups
Abacus	play-money	scale

Number Board

1 - 50

1 2 3 4 5 6 7 8
9 10

11 12 13 **14** 15 16 17
18 19 20

21 22 **23** **24** 25 26 27
28 29 30

31 32 **33** **34** 35 36 37
38 39 40

41 42 **43** **44** 45 46 47
48 49 50

Number Board   (51 – 100)

51 52 53 **54** 55   56 57 **58**
59 60

61 62 63 **64** 65 66 67 **68**
69 70

71 72 **73** **74** 75 76 77 **78**
79 80

81 82 **83** **84** 85 86 87 **88**
89 90

91 92 93 94 85 96 97 98 99
100

Count and write

Can you name these fruits?   Count and write the number 1 in each box.

Mango

apple

pineapple

lime

How Many?

Count objects and write the correct number in the box

Draw and colour your fruits

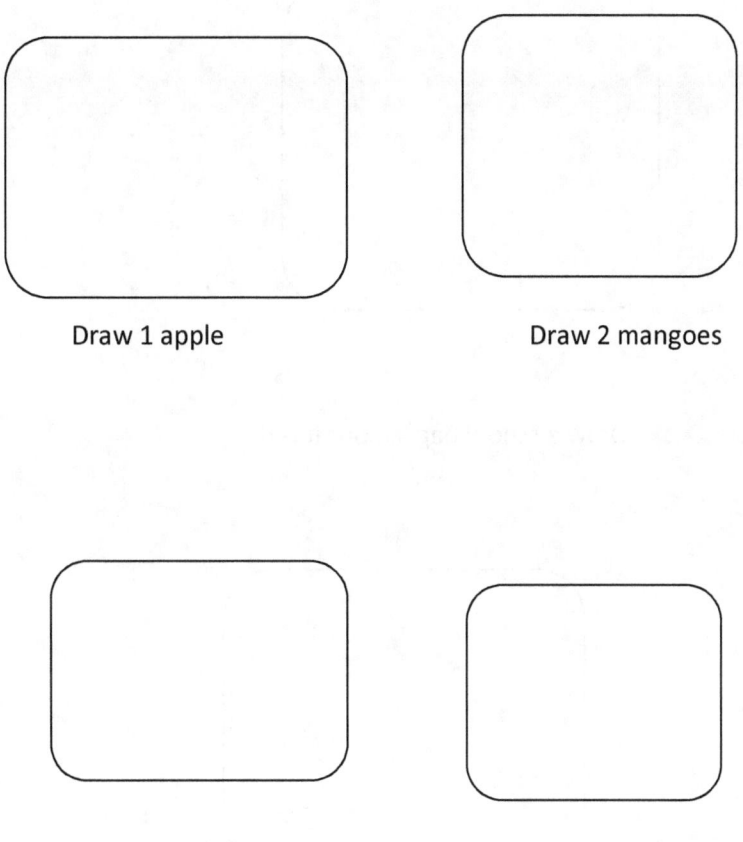

Draw 1 apple

Draw 2 mangoes

Draw 2 bananas

Draw 1 sweetsop

Colour Me

Draw a school bag, colour it red

Draw a crayon, colour it yellow

Draw a lunch – kit. Colour it blue.

Colour each shape correctly

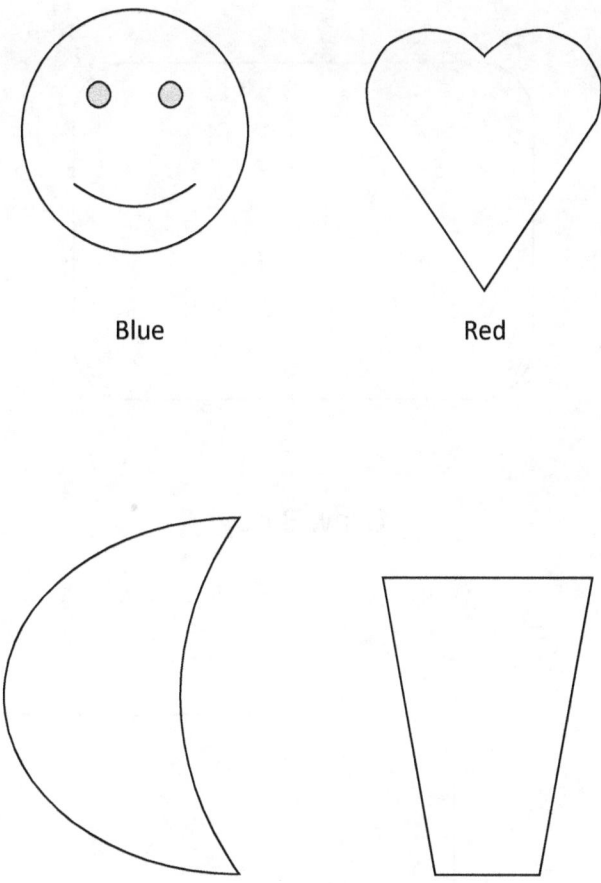

Blue

Red

Yellow

Green

Counting Animals

Can you count to 3?

Show 3 fingers

Draw 3 dogs

Draw 3 cats

Name the Shapes

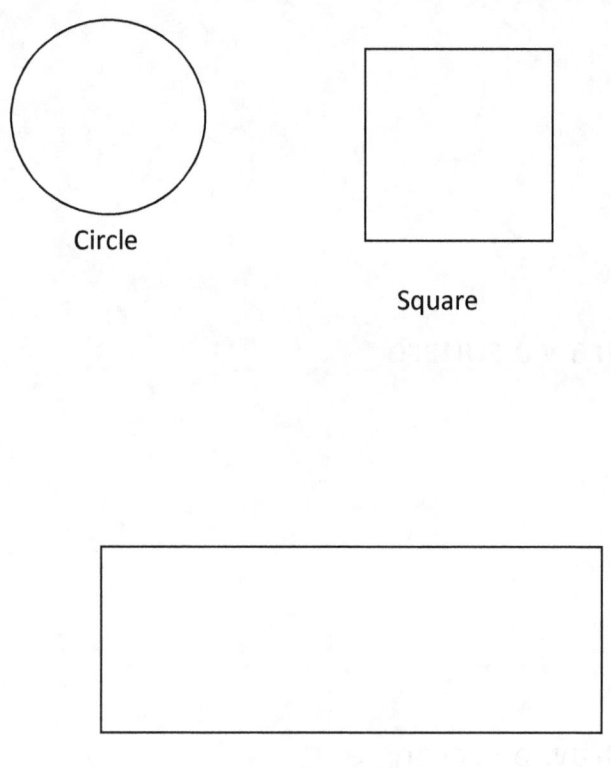

Circle

Square

Rectangle

Draw Shapes

Draw a circle

Draw a square

Draw a rectangle

Match and Colour Shapes

Join the dots to match shapes

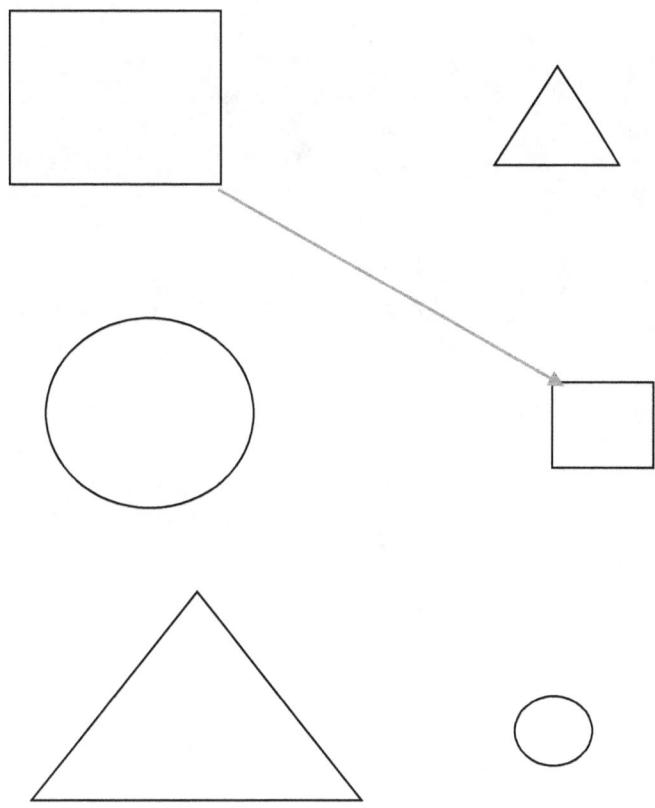

Let us Count to 50

1 2 3 4 5 6 7 8 9 10

11 12 13 14 15 16 17 18
19 20

21 22 23 24 25 26 27
28 29 30

31 32 33 34 35 36 37 38
39 40

41 42 43 44 45 46 47 48
49 50

The Months of the Year

Look at a Calendar. Let us name the months

Can you count them?   When is your birthday?

Can you find it on the calendar?

How old are you?

What year are we in now?

Date of Birth: _____

_____   _____   _____

Day   Month   Year

Number 4

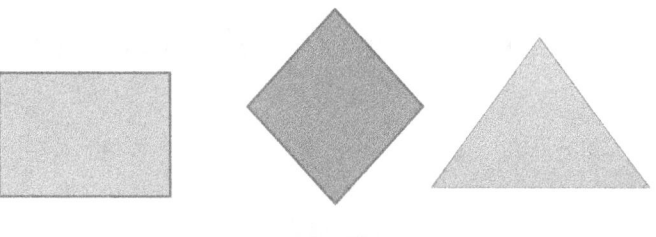

3 + 1 = ☐   1 + 3 = ☐

2 + 2 = ☐

Count to 4

Draw the correct number of books and cups then write the answers

3 books   +   1 book = ⬜ books

2   cups   +   2   cups = ⬜ cups

Here comes 5

Count   Colour   Write

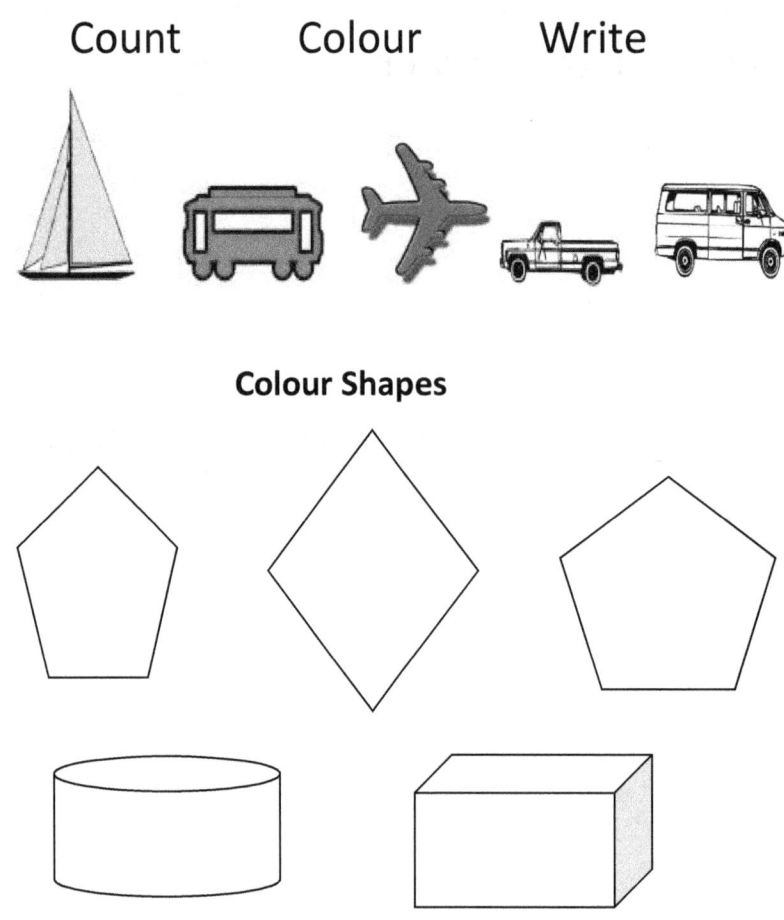

Colour Shapes

Write the number 5 on the line ____

Words and Numbers   1 - 10

1	One
2	Two
3	Three
4	Four
5	Five
6	Six
7	Seven
8	Eight
9	Nine
10	Ten

Match Numbers to Words

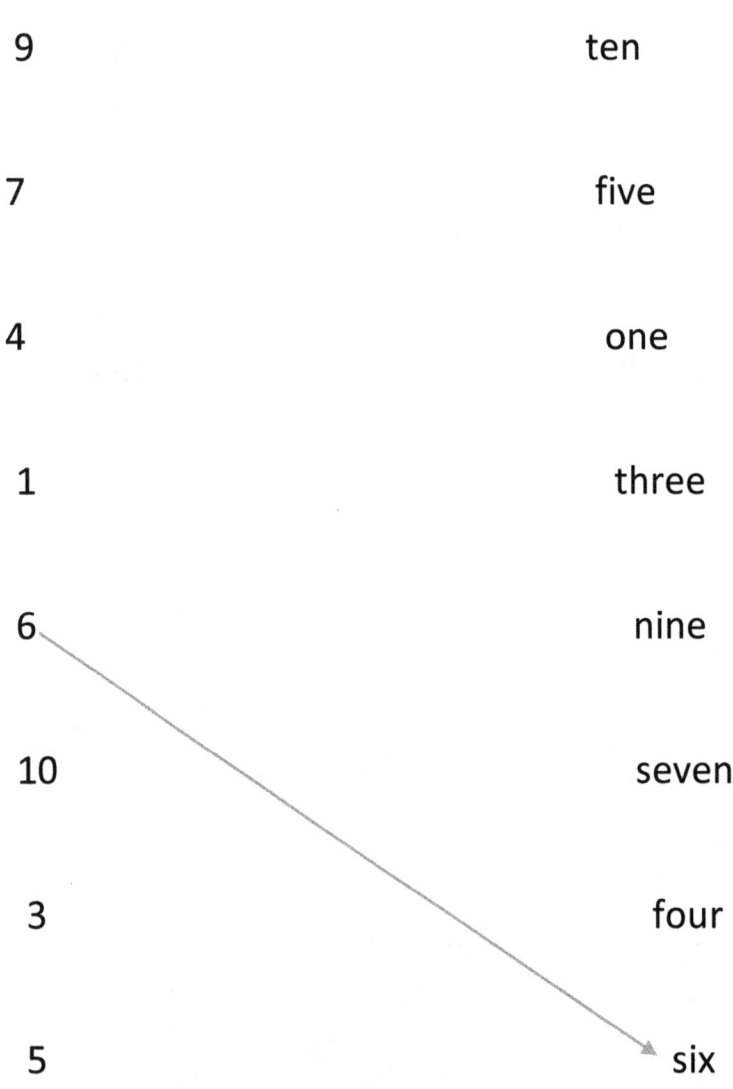

9   ten

7   five

4   one

1   three

6   nine

10   seven

3   four

5   six

Match and Draw

2   eight

8   two

Draw 5 balloons and colour them

Yeah for 6

1   2   3   4   5   6

Draw and colour 6 marbles

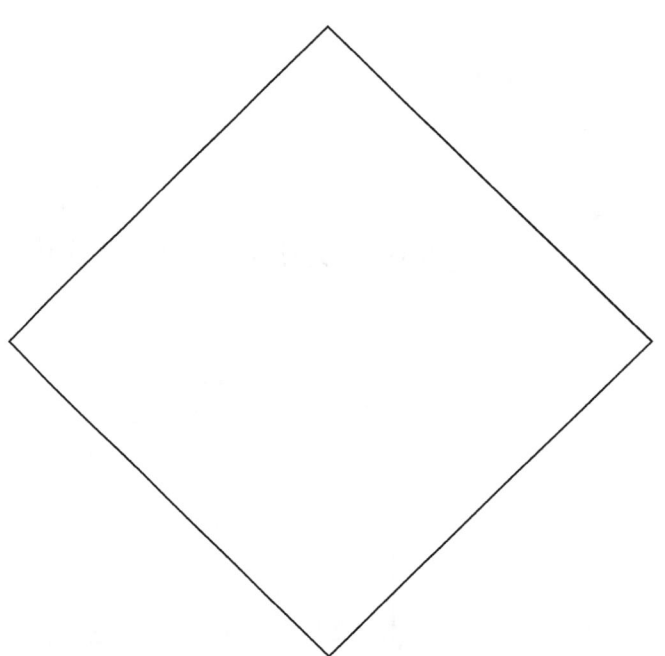

Six and 6

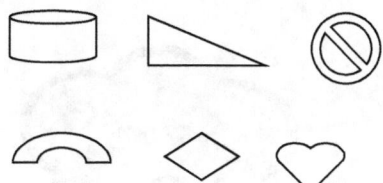

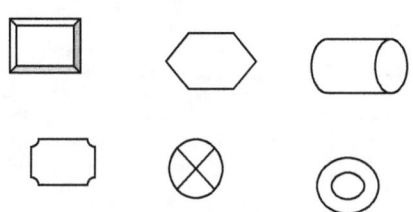

3 + 3 =   4 + 2 =

5 + 1 =

How many apples are on the tree?

Clap Stamp Snap

Clap your hand for number 1

Stamp your feet for number 2

Snap your fingers for number 3

Stamp your feet for number 4

Clap your hands for number 5

Mrs 7

Mrs. 7 Mrs 7 where are you?

Across the way come straight

This way

There you are Mrs 7

What is your favourite ice cream flavour?

Draw 7 ice cream cones.

Mr 8 by the Gate

Meet me by the gate Mr Eight

Count to 8

Show 8 fingers

Count

Now Add

Use your counters to find the answers

4 + 4 = ☐   3 + 5 = ☐

6 + 2 = ☐   7 + 1 = ☐

2 + 6 = ☐   1 + 7 = ☐

5 + 3 = ☐   8 + 0 = ☐

It is empty

Is there anything in this set?

This set is empty.

Write 0 in the box.

The Empty Set

Draw an empty set.

Write 0 below it.

Hey Mr. 9

Colour and count. Use pencil crayons.

Write the number 9 on the line __

1   2   3   4   5   6 7 8   9

Use your counters to find the answers

2 + 7 = ☐   7 + 2 = ☐

3 + 6 = ☐   6 + 3 = ☐

4 + 5 = ☐   5 + 4 = ☐

1 + 8 = ☐   8 + 1 = ☐

9 + 0 = ☐   0 + 9 = ☐

Count to Ten

1 2 3 4 5 6 7 8 9 10

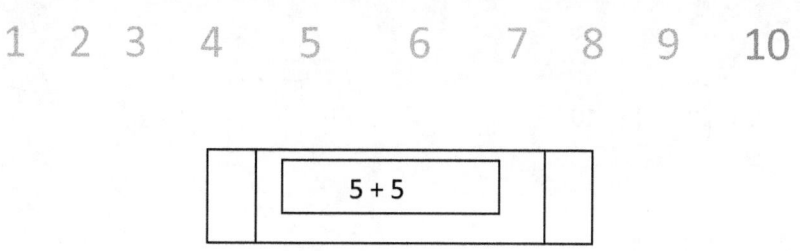

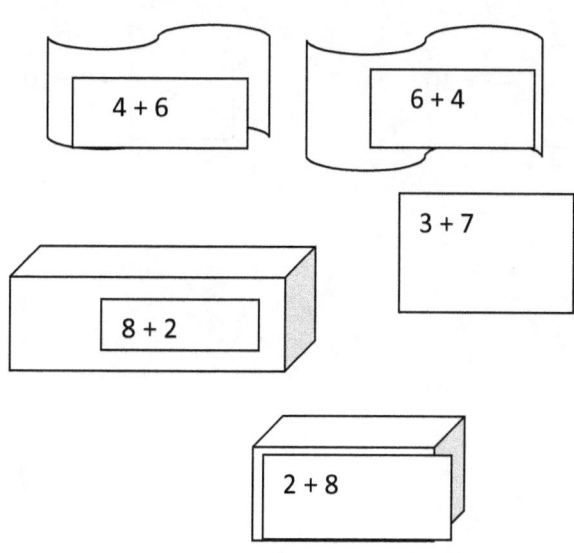

35

Find the missing numbers.

$6 + \boxed{} = 10$ $2 + \boxed{} = 10$

$5 + \boxed{} = 10$ $7 + \boxed{} = 10$

$\boxed{} + 9 = 10$ $\boxed{} + 8 = 10$

$\boxed{} + 3 = 10$ $\boxed{} + 9 = 10$

Ten to Music

Songs: 1.Ten Little Indians

2. Ten little fingers

Hit your triangles 10 times

Hit your drum ten times

Shake your shakers 10 times

Tap your desk/table 10 times

Now count 10 items and put on your desk.

Let's count by 10

10   20   30   40   50

60   70   80   90   100

Let Us Play Bingo

B	I	N	G	O
3	7	4	2	11
9	15	1	6	8
12	18	13	17	20
19	16	15	0	14

What time is it?

It is time to get up

It is time for breakfast

It is time for school

Take Away

Put 5 blocks on your desk. Take up 1.

How many blocks are left on your desk?

Count them.

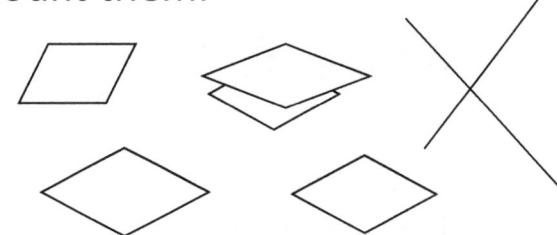

5 blocks take away I block = ☐ blocks

Now put 3 counters on your desk.

Take up 2. How many are left?

3 − 2 = ☐

Fun with Left and Right

Raise your right hand.

Raise your left hand

Turn your body to the right

Turn your body to the left

Colour the ball to your left

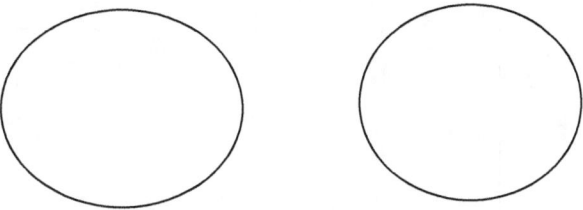

Draw your favourite fruit to the right of this page.

Sing and dance to "Hokey Pokey" Song.

Congratulations!

You have completed your first

Number Play Book.

Draw a picture of yourself below

Be sure to smile

My name is ------------------------

I have completed my first
Number Play Book

Number Play Yeah!

Parent/Teacher Instruction Notes

Before introducing your child to the exercises in _Number Play Yeah_ kindly source items for the Math Lab. This is necessary as he/she will be embarking on a scientific journey. He will only say YEAH! To Mathematics if he is given the tools to explore and make discoveries. The fun begins at the lab and the next step is to do the exercises which follow. Always read the instructions for the child.

"Number Play Yeah" is designed to stimulate your child to explore, therefore from time to time he must visit the Math Lab to discover answers. Find a suitable spot to display items and supervise him as he explores. You may not have all items right now but start and build as you go.

Math Lab

Have the following items available to enhance 'Number Play'

Large bottle tops, pencils, jumbo crayons, pencil crayons, water paint and paint brushes, musical items such as drums, triangles, shakers, colour chart, drinking straws, coloured fudge sticks.

Blocks of different colours and sizes, solid shapes, cut out shapes (triangle, square, rectangle, circle and oval

Metre stick, rulers, number chart, pictures of animals, fruits etc. for counting and discussions. Puzzles, toy cash register, shopping list. Supermarket chart, menu board, number cut- outs Lego blocks. Pictures of children at play for discussion, deck of cards and number games.

Feel free to add items which are age appropriate and relevant to the Math Lab.

Number Art

Guide your children in creating beautiful art work as they identify numbers.

<u>Material</u>: sandbox, paint and paint brushes, glue, basin with water, number cards (numbers written in bold print).

Assist children to pour glue onto each number following the formation. Sprinkle sand on glue and allow to dry. Paint sand with colour of choice. Put design on display and encourage children to discuss colours and numbers.

Source number and colour rhymes and song for young children on You Tube. Invite children to sing along.

Use crayons and paint to create number pattern on news print and cards.

Parent/ Teacher Instructional Notes

In order to effectively complete the exercises in *Number Play Yeah,* children need to use material from the Math Lab to assist them in finding answers. Kindly ensure that children have their tools from the <u>*Math Lab*</u> to learn number concepts.

<u>Number Board</u>

Introduce your child to the Number Board at the beginning of the book. Help him to count and show him how to consult his board when necessary. Although Number Play has its focus on numbers 1- 10 children need to be aware of other numbers and their relationship with other areas of learning.

Count and Write

Preparation:

Before introducing the child to page 6 place a number of items on the table. Ask him to bring you one of each e.g. 1book, 1bag, 1cup, 1orange. Ask him to show you 1 finger.

Read instructions and allow child to attempt exercises. Assist where necessary.

How Many?

Help child to count to 2 by counting objects around the house e.g. 2cups, two shoes and 2 chairs.

Have child name fruits and talk about his favourite fruit.

Ask the child to identify the number 2 on the Counting Board. Read the instructions and supervise the child.

Draw and colour your fruits

Ask the child to name fruits. Read instructions

Colour Me & Colour Each Shape

Preparation: Watch you tube video - Electric colours Busy Beavers sing along- red green yellow blue.

Source bits of fabric with colours red, yellow, blue, green, orange purple to use as necessary. For colours red, yellow blue, invite child to join you in singing colour rhyme.

Colour Rhyme

Red yellow blue, red yellow blue

These are the colours for you

Red yellow blue, colours for you

For you, for you, for you

Red yellow blue.

Let child wave colour fabric as named while saying rhyme.

Teach child the following shape rhymes when introducing shapes.

Triangle Three Rhyme

Three sides meet three corners (x3)

My you are so sharp

Pointed, pointed, pointed

Triangle, triangle, triangle

Three sides meet three corners

If you have a musical triangle this would be a good time to have child play while singing triangle rhyme. Show a triangle cut out or picture of triangle.

Circle Round Rhyme

Here we go around and around (x2)

Around and around we go

Forming the big round (x2)

Forming the big round circle

Around and around, around and around

That's the circle for me.

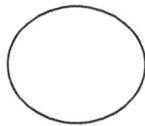

Instruct child to

Use playdough to form circle

Use crayons to make circle shapes on newsprint

Make circles in sandbox

Counting Animals

Invite your child to practise writing number 3 in the sand and also do shadow writing. Prepare an outline of number 3 on newsprint and let him use pencil crayons to join the dots. Ask him to name other animals.

Square 4 Rhyme

A four sided figure

With four corners squared

No turning no twisting for me

No slanting no bending

No leaning or shaking

As square as a square can be

My sides stand like soldiers on duty

Protecting the rest of my form

I'm shaped like a box

Size evenly matched

As square as a square can be

You sure are a square

As square as a square can be

No twisting no turning

No slanting or bending

That's how a square should be

Rectangle Rhyme

Guess this if you can

Four sides but not all the same

Two sides long two sides short

What shape is that?

That will be that will be

That will be a rectangle

Let's Count to 50

Your child should have been introduced to the Counting Board before. He is excited about learning new numbers. Challenging him to count to 50. Let him collect counting blocks from the Math Lab. Count with him more than once. Invite him to point to each number on his page and count. Assist where necessary.

Praise him for his effort.

The Months of the Year

Let your child observe the calendar. Point out the months and numbers. Show him the month and date on which he/she was born. Assist him to write his date of birth. Write it on a card and let him copy it to complete the exercise. Let him count months. Discuss birthday.

Count to 4

Help child to sky write number 4 while singing this jingle. *Come straight down across to the right*

Strike in the middle and that makes four (x2)

Number 4

Get your child to use drinking straws, playdough and fudge sticks to form number 4.

Here Comes 5

Prepare sandbox, bits of coloured wool, playdough, learn the 5 Rhyme,

Invite your child to identify number 5 on the Number Board and write it in the sand, use playdough and wool to design the number 5. Sing rhyme along with your child while sky writing number 5.

Across and down then curve to the right

Across and down then curve to the right

Across and down then curve to the right

Now say hello to five.

Have child count 5 blocks and place in a basket. Show five fingers. Read instructions for child.

Words and Numbers

Preparation: Use coloured cartridge paper and markers to prepare number and word cards 1 – 10. Help your child to identify words and numbers by matching number **one** word card to numeral **1.**   Two - 2 etc.

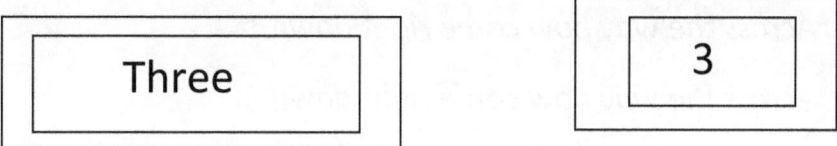

Say numbers and have child identify words and numbers.

Read instruction for child.

Yeah for 6

 Prepare a large card with the number 6 printed on it. Have sand, glue paint and paint brushes. Assist child in putting glue onto the number and sprinkle sand on glue. Allow glue to dry then let the child paint the sand.   Read instructions.

6 and six

Have child count to 6 and find number 6 on the counting board. Discuss the exercises on the given pages. Read instructions.

Clap Stamp Snap

Prepare number cards displaying word on one side and number on the next.

Instruct child to clap, snap fingers or stamp feet when a particular number is call.

Show number cards and give instruction e.g. snap for 2, stamp for 1 and clap for 3.

 Show your child how to have fun with numbers. Enjoy the exercise with your child.

Mrs. 7

Help child to practise shadow writing while saying rhyme

Mrs. 7 Mrs 7 where are you?

Across the way now come right down

Across the way now come right down

Across the way …. Now come … right … down

There you are oh dear you are

There you are Mrs. 7.

Mr. 8

Help your child to use glue to form the number 8 then sprinkle sand on it. Use wool and play dough to form number 8.

Engage child in singing the Mr. 8 Rhyme.

<u>Meet me by the Gate Mr. 8</u>

Meet me by the gate Mr. 8

Meet me by the gate Mr. 8

Right behind 7 Mr 8 Mr. 8

Meet me by the gate Mr. 8

Draw Mr. 8 by the gate

Discuss exercises and read instructions.

Now Add

Provide 2 plastic bowls and 8 coloured counters

Guide child in putting given number of counters in bowls then counting to find answers e.g. 2 blocks and 6 blocks = ☐ blocks.

It's Empty

Put 3 plates on the table. Plate A has 2 cookies. Plate B is empty Plate C an apple. Ask child: What is in Plate A? How many?

Is there anything in Plate B? What can you say about Plate B?

What is in Plate C?

Ask child to use his/her fingers to show number of cookies and apples in Plate A and C.

What number do you use for the plate that does not have anything in it?   Discuss.

Introduce the number 0. Guide him in completing the exercise. Write the word zero on a card and show him.

Hey Mr. 9

Assist child in using Number Board and counters to count to 9.

Instruct child to tap table 9 times and stamp feet nine times.

Outline the number 9 on a card and ask child to connect dots.

Beat drum nine times, hit triangle 9 times

Sing rhyme and invite the child to join in.

Rhyme for Mr. 9

Where Mr. 9 is oh where is Mr. 9?

Between 8 and 10 between 8 and 10

Right after 8 and just before 10

Search high and low, search near and far

Right after 8 and just before 10

That's where you'll find Mr.9

Read instructions and allow child to attempt exercises

Here Comes 10

If you have access to the internet visit utube channel for Number 10 songs and rhymes for young children

Help your child to count to 10 using Counting Board and counters. Help children to create designs around the number 10 on cards.

Let us Play Bingo

Design a Bingo Card. Have coloured markers. Explain rule of the game and read numbers. The first child to cover numbers which spell BINGO is the winner. Give a prize if possible.

What time is it?

Prepare a clock made of cardboard or other suitable material for a young child. Assist child in showing given time on his clock e.g. What time do you get up in the mornings?

Can you show that time on your clock? Give assistance where necessary.

Discuss time for breakfast, time for school and bed time. Guide child in setting time on the hour.

Take Away

Use counters and other items to show child how to take away.

Give instructions for child to apply subtraction rules e.g.

If you have 4 balloons and give your brother 1 how many balloons will you have left?

Show him/her how to use counters to find answers to questions

Discuss the exercises and allow him to attempt them.

Fun with Left and Right

Your son /daughter has come to the end of his/her first Number Play book. Kindly source the words of the song "Hokey Pokey" so that both of you can complete *Number Play Yeah* with a song and a dance. Hope your child or student had fun. Look out for "Number Play Hurrah!

Congratulate your child or student for successfully completing

Number Play Yeah!

Thank you for taking the time to play. Cheers!

For more exciting Number Play Exercises get your copy of *Number Play Hurrah!* And *Number Play This Way.*

Visit:

www.amazon.com/author/www.parentingnmore.com

Other books by Colleen M. Thompson

When Boys Don't Read

Forgiveness: Tough Choice Great Benefits

How to Motivate Your Son to Achieve

Success in Reading

Lemonade: A Second Serving

Visit:

www.amazon.com/author/www.parentingnmore.com

Blog: www.parentingnmore.com

Colleen M. Thompson

Published July 2015

References

www.khanacademy.org/early-math

www.khanacademy.org/youc

www.starfall.com